The Blood of the Nation

The Blood of the Nation

A STUDY OF THE DECAY OF RACES
THROUGH THE SURVIVAL OF THE UNFIT

BY

DAVID STARR JORDAN

President of Leland Stanford Jr. University

**University Press of the Pacific
Honolulu, Hawaii**

The Blood of the Nation:
A Study of the Decay of Races Through the Survival of the Unfit

by
David Starr Jordan

ISBN: 1-4102-0953-9

Copyright © 2003 by University Press of the Pacific

Reprinted from the 1902 edition

University Press of the Pacific
Honolulu, Hawaii
http://www.universitypressofthepacific.com

All rights reserved, including the right to reproduce this book, or portions thereof, in any form.

In order to make original editions of historical works available to scholars at an economical price, this facsimile of the original edition of 1902 is reproduced from the best available copy and has been digitally enhanced to improve legibility, but the text remains unaltered to retain historical authenticity.

I
IN PEACE

THE BLOOD OF THE NATION.

I

In Peace.

"Over trench and clod
 Where we left the bravest of us,
 There's a deeper green of the sod."
                    H. H. Brownell.

In this paper I shall set forth two propositions: the one self-evident; the other not apparent at first sight, but equally demonstrable. *The blood of a nation determines its history.* This is the first proposition. The second is, *The history of a nation determines its blood.* As for the first, no one doubts that the character of men controls their deeds. In the long run and with masses of mankind this must be true, however great the emphasis we may lay on individual initiative or on individual variation.

Equally true is it that the present character of a nation is made by its past history. Those who are alive to-day are the resultants of the stream of heredity as modified by the vicissitudes through which the nation has passed. The blood of the nation flows in the veins of those who survive. Those who die without descendants can not color the stream of heredity. It must take its traits from the actual parentage.

The word "blood" in this sense is figurative only, an expression formed to cover the qualities of heredity. Such traits, as the phrase goes, "run in the blood." In the earlier philosophy it was held that blood was the actual physical vehicle of heredity, that the traits bequeathed from sire to son as the characteristics of families or races ran literally in the literal blood. We know

In Peace

now that this is not the case. We know that the actual blood in the actual veins plays no part in heredity, that the transfusion of blood means no more than the transposition of food, and that the physical basis of the phenomena of inheritance is found in the structure of the germ cell and its contained germ-plasm.

But the old word well serves our purposes. The blood which is "thicker than water" is the symbol of race unity. In this sense the blood of the people concerned is at once the cause and the result of the deeds recorded in their history. For example, wherever an Englishman goes, he carries with him the elements of English history. It is a British deed which he does, British history that he makes. Thus, too, a Jew is a Jew in all ages and climes, and his deeds everywhere bear the stamp of Jewish individuality. A Greek is a Greek; a Chinaman remains a Chinaman. In like fashion the race traits color all history made by Tartars, or negroes, or Malays.

The climate which surrounds a tribe of men may affect the activities of these men as individuals or as an aggregate, education may intensify their powers or mellow their prejudices, oppression may make them servile or dominion make them overbearing; but these traits and their resultants, so far as science knows, do not "run in the blood," they are not "bred in the bone." Older than climate or training or experience are the traits of heredity, and in the long run it is always "blood which tells."

On the other hand, the deeds of a race of men must in turn determine its blood. Could we with full knowledge sum up the events of the past history of any body of men, we could indicate the kinds of men destroyed in these events. The others would be left to

In Peace

write the history of the future. It is the "man who is left" in the march of history who gives to history its future trend. By the "man who is left" we mean simply the man who remains at home to become the father of the family as distinguished from the man who in one way or another is sacrificed for the nation's weal or woe. If any class of men be destroyed by political or social forces or by the action of institutions, they leave no offspring, and their like will cease to appear.

"Send forth the best ye breed." This is Kipling's cynical advice to a nation which happily can never follow it. But could it be accepted literally and completely, the nation in time would breed only second-rate men. By the sacrifice of their best or the emigration of the best, and by such influences alone, have races fallen from first-rate to second-rate in the march of history.

The Blood of the Nation

For a race of men or a herd of cattle are governed by the same laws of selection. Those who survive inherit the traits of their own actual ancestry. In the herd of cattle, to destroy the strongest bulls, the fairest cows, the most promising calves, is to allow those not strong nor fair nor promising to become the parents of the coming herd. Under this influence the herd will deteriorate, although the individuals of the inferior herd are no worse than their own actual parents. Such a process is called race-degeneration, and it is the only race-degeneration known in the history of cattle or men. The scrawny, lean, infertile herd is the natural offspring of the same type of parents. On the other hand, if we sell or destroy the rough, lean, or feeble calves, we shall have a herd descended from the best. It is said that when the short-horned Durham cattle first attracted attention in England, the long-horns which preceded them, inferior for beef or milk, vanished "as if smitten by a pestilence." The fact was that, being less valuable, their owners chose to destroy them rather than the finer Durhams. Thus the new stock came from the better Durham parentage. If conditions should ever be reversed and the Durhams were chosen for destruction, then the long-horns might again appear, swelling in numbers as if by magic, unless all traces of the breed had in the meantime been annihilated.

In selective breeding with any domesticated animal or plant, it is possible, with a little attention, to produce wonderful changes for the better. Almost anything may be accomplished with time and patience. To select for posterity those individuals which best meet our needs or please our fancy, and to destroy those with unfavorable qualities, is the function of artificial selection. Add to this the occasional crossing of unlike forms to promote new and desirable variations, and we have the whole secret of selective breeding. This process Youatt calls the "magician's wand" by which man may summon up and bring into existence any form of animal or plant useful to him or pleasing to his fancy.

In the animal world, progress comes mainly through selection, natural or artificial, the survival of the fittest to become the parent of the new generation. In the world of man similar causes produce similar results. The word "progress" is, however, used with a double meaning, including the advance of civilization as well as race improvement. The first of these meanings is entirely distinct from the other. The results of training and education lie outside the scope of the present discus-

In Peace

sion. By training the force of the individual man is increased. Education gives him access to the accumulated stores of wisdom built up from the experience of ages. The trained man is placed in a class relatively higher than the one to which he would belong on the score of heredity alone. Heredity carries with it possibilities for effectiveness. Training makes these possibilities actual. Civilization has been defined as "the sum total of those agencies and conditions by which a race may advance independently of heredity." But while education and civilization may greatly change the life of individuals, and through them that of the nation, these influences are spent on the individual and the social system of which he is a part. So far as science knows, education and training play no part in heredity. The change in the blood which is the essence of race-progress, as distinguished from progress in civilization, finds its cause in selection only.

To apply to nations the principles known to be valid in cattle-breeding, we may take a concrete example, that of the alleged decadence of France. It is claimed that the birth-rate is falling off in France, that the stature is lower, and the physical force less among the French peasantry than it was a century ago. If all this is true, then the cause for it must be in some feature of the life of France which has changed the normal processes of selection.

In the present paper I shall not attempt to prove these statements. They rest, so far as I know, entirely on assertions of French writers, and statistics are not easily obtained. It suffices that an official commission has investigated the causes of reduced fertility, with chiefly negative results. It is not due

In Peace

primarily to intemperance nor vice nor prudence nor misdirected education, the rush to "ready-made careers," but to inherited deficiencies of the people themselves. It is not a matter of the cities alone, but of the whole body of French peasantry. Legoyt, in his study of "the alleged degeneration of the French people," tells us that "it will take long periods of peace and plenty before France can recover the tall statures mowed down in the wars of the republic and the First Empire," though how plenty can provide for the survival of the tallest this writer does not explain. Peace and plenty may preserve, but they can not restore.

It is claimed, on authority which I have failed to verify, that the French soldier of to-day is nearly two inches shorter than the soldier of a century ago. One of the most important of recent French books, by Edmond Demolins, asks, "In what consists the superiority of the Anglo-Saxon?" The answer is found in defects of training and of civic and personal ideals, but the real cause lies deeper than all this. Low ideals in education are developed by inferior men. Dr. Nordau and his school of exponents of "hand-painted science" find France a nation of decadents, — a condition due to the inherited strain of an overwrought civilization. With them the word "degenerate" is found adequate to explain all eccentricities of French literature, art, politics, or jurisprudence.

But science knows no such things as nerve-stress inheritance. If it did, the peasantry of France have not been subjected to it. Their life is hard, no doubt, but not stressful; and they suffer more from nerve-sluggishness than from any form of enforced psychical activity. The kind of degeneration

In Peace

Nordau pictures is not a matter of heredity. When not simply personal eccentricity, it is a phase of personal decay. It finds its causes in bad habits, bad training, bad morals, or in the desire to catch public attention for personal advantage. It has no permanence in the blood of the race. The presence on the Paris boulevards of a mob of crazy painters, maudlin musicians, drunken poets, and sensation-mongers proves nothing as to race degeneracy. When the fashion changes, they will change also. Already the fad of "strenuous life" is blowing them away. Any man of any race withers in an atmosphere of vice, absinthe, and opium. The presence of such an atmosphere may be an effect of race decadence, but it is not a cause of the lowered tone of the nation.

Evil influences may kill the individual, but they cannot tarnish the

The Blood of the Nation

stream of heredity. The child of each generation is free-born so far as heredity goes, and the sins of the fathers are not visited upon him. If vice strikes deeply enough to wreck the man, it is likely to wreck or kill the child as well, not through heredity, but through lack of nutrition. The child depends on its parents for its early vitality, its constitutional strength, the momentum of its life, if we may use the term. For this a sound parentage demands a sound body. The unsound parentage yields the withered branches, the lineage which speedily comes to the end. But this class of influences, affecting not the germ-plasm, but general vitality, has no relation to hereditary qualities, so far as we know.

In heredity there can be no tendency downward or upward. Nature repeats, and that is all. From the actual parents actual qualities are received, the

In Peace

traits of the man or woman as they might have been, without regard, so far as we know, to the way in which these qualities have been actually developed.

The evolution of a race is selective only, never collective. Collective evolution, the movement upward or downward of a people as a whole, irrespective of education or of selection, is, as Lepouge has pointed out, a thing unknown. "It exists in rhetoric, not in truth nor in history."

No race as a whole can be made up of "degenerate sons of noble sires." Where decadence exists, the noble sires have perished, either through evil influences, as in the slums of great cities, or else through the movements of history or the growth of institutions. If a nation sends forth the best it breeds to destruction, the second best will take their vacant places. The weak, the vicious, the unthrifty will propagate, and in default of better will have the land to themselves.

We may now see the true significance of the "Man of the Hoe," as painted by Millet and as pictured in Edwin Markham's verse. This is the Norman peasant, low-browed, heavy-jawed, "the brother of the ox," gazing with lacklustre eye on the things about him. To a certain extent, he is typical of the French peasantry. Every one who has travelled in France knows well his kind. If it should be that his kind is increasing, it is because his betters are not. It is not that his back is bent by centuries of toil. He was not born oppressed. Heredity carries over not oppression, but those qualities of mind and heart which invite or which defy oppression. The tyrant harms those only that he can reach. The new generation is free-born, and slips from his hands, unless its traits be of the kind which demand new tyrants.

In Peace

Millet's "Man of the Hoe" is not the product of oppression. He is primitive, aboriginal. His lineage has always been that of the clown and swineherd. The heavy jaw and slanting forehead can be found in the oldest mounds and tombs of France. The skulls of Engis and Neanderthal were typical men of the hoe, and through the days of the Gauls and Romans the race was not extinct. The "lords and masters of the earth" can prove an *alibi* when accused of the fashioning of the terrible shape of this primitive man. And men of this shape persist to-day in regions never invaded by our social or political tyranny, and their kind is older than any existing social order.

That he is "chained to the wheel of labor" is the result, not the cause, of his impotence. In dealing with him, therefore, we are far from the "labor problem" of to-day, far from the workman brutalized by machinery, and from all the wrongs of the poor set forth in the conventional literature of sympathy.

In our discussion of decadence we turn to France first simply as a convenient illustration. Her sins have not been greater than those of other lands, nor is the penalty more significant. Her case rises to our hand to illustrate a principle which applies to all human history and to all history of groups of animals and plants as well. Our picture, such as it is, we must paint with a broad brush, for we have no space for exceptions and qualifications, which, at the most, could only prove the rule. To weigh statistics is impossible, for the statistics we need have never been collected. The evil effects of "military selection" and allied causes have been long recognized by students of social science, but their

In Peace

ideas have not penetrated into the common literature of common life.

The survival of the fittest in the struggle for existence is the primal cause of race-progress and race-changes. But in the red field of human history the natural process of selection is often reversed. The survival of the unfittest is the primal cause of the downfall of nations. Let us see in what ways this cause has operated in the history of France.

First, we may consider the relation of the nobility to the peasantry, the second to the third estate.

The feudal nobility of each nation was in the beginning made up of the fair, the brave, and the strong. By their courage and strength their men became the rulers of the people, and by the same token they chose the beauty of the realm to be their own.

In the polity of England this superiority was emphasized by the law of primogeniture. On "inequality before the law" British polity has always rested. Men have tried to take a certain few, to feed these on "royal jelly," as the young queen bee is fed, and thus to raise them to a higher class, distinct from all the workers. To take this leisure class out of the struggle and competition of life, so goes the theory, is to make of the first-born and his kind harmonious and perfect men and women, fit to lead and control the social and political life of the state. In England the eldest son is chosen for this purpose,— a good arrangement, according to Samuel Johnson, "because it insures only one fool in the family." For the theory of the leisure class forgets that men are made virile by effort and resistance, and the lord developed by the use of "royal jelly" has rarely been distinguished by perfection of manhood.

In Peace

The gain of primogeniture came in the fact that the younger sons and the daughters' sons were forced constantly back into the mass of the people. Among the people at large this stronger blood became the dominant strain. The Englishmen of to-day are the sons of the old nobility, and in the stress of natural selection they have crowded out the children of the swineherd and the slave. The evil of primogeniture has furnished its own antidote. It has begotten democracy. The younger sons in Cromwell's ranks asked on their battle-flags why the eldest should receive all and they nothing. Richard Rumbold, whom they slew in the Bloody Assizes, "could never believe that Providence had sent into the world a few men already booted and spurred, with countless millions already saddled and bridled for these few to ride." Thus these younger sons became the

The Blood of the Nation

Roundhead, the Puritan, the Pilgrim. They swelled Cromwell's army, they knelt at Marston Moor, they manned the "Mayflower," and in each generation they have fought for liberty in England and in the United States. Studies in genealogy show that all this is literally true. All the old families in New England and Virginia trace their lines back to nobility, and thence to royalty. Almost every Anglo-American has, if he knew it, noble and royal blood in his veins. The Massachusetts farmer, whose fathers came from Plymouth in Devon, has as much of the blood of the Plantagenets, of William and of Alfred, as flows in any royal veins in Europe. But his ancestral line passes through the working and fighting younger son, not through him who was first born to the purple. The persistence of the strong shows itself in the prevalence of the leading qualities of her dominant strains of blood, and it is well for England that her gentle blood flows in all her ranks and in all her classes. When we consider with Demolins "what constitutes the superiority of the Anglo-Saxon," we shall find his descent from the old nobility, "Saxon and Norman and Dane," not the least of its factors.

On the continent of Europe the law of primogeniture existed in less force, and the results were very distinct. All of noble blood were continuously noble. All belonged to the leisure class. All were held on the backs of a third estate, men of weaker heredity, beaten lower into the dust by the weight of an ever-increasing body of nobility. The blood of the strong rarely mingled with that of the clown. The noblemen were brought up in indolence and ineffectiveness. The evils of dissipation wasted their individual lives, while casting an ever-increasing burden on the villager and on the "farmer who must pay for all."

Hence in France the burden of taxation led to the Revolution and its Reign of Terror. I need not go over the details of dissipation, intrigue, extortion, and vengeance which brought to sacrifice the "best that the nation could bring." In spite of their lust and cruelty, the victims of the Reign of Terror were literally the best from the standpoint of race development. Their weaknesses were those of training in luxury and irresponsible power. These effects were individual only; and their children were free-born, with the capacity to grow up truly noble if removed from the evil surroundings of the palace.

In Thackeray's "Chronicle of the Drum," the old drummer, Pierre, tells us that

In Peace

"Those glorious days of September
　　Saw many aristocrats fall;
'Twas then that our pikes drank the blood
　　In the beautiful breast of Lamballe.

"Pardi, 'twas a beautiful lady,
　　I seldom have looked on her like;
And I drummed for a gallant procession
　　That marched with her head on a pike."

Then they showed her pale face to the Queen, who fell fainting; and the mob called for her head and the head of the King. And the slaughter went on until the man on horseback came, and the mob, "alive but most reluctant," was itself forced into the graves it had dug for others.

And since that day the "best that the nation could bring" have been without descendants, the men less manly than the sons of the Girondins would have been, the women less beautiful than the daughters of Lamballe. The political changes which arose may have been for the better; the change in the blood was all for the worse.

Other influences which destroyed the best were social repression, religious intolerance, and the intolerance of irreligion and unscience. It was the atheist mob of Paris which destroyed Lavoisier, with the sneer that the new republic of reason had no use for savants. The old conservatism burned the heretic at the stake, banished the Huguenot, destroyed the lover of freedom, silenced the agitator. Its intolerance gave Cuvier and Agassiz to Switzerland, sent the Le Contes to America, the Jouberts to Holland, and furnished the backbone of the fierce democracy of the Transvaal. While not all agitators are sane, and not all heretics right-minded, yet no nation can spare from its numbers those men who think for themselves and those who act for themselves. It cannot afford to drive away or destroy those who are filled with religious zeal, nor those whose religious zeal takes a form not approved by tradition nor by consent of the masses. All movements toward social and religious reform are signs of individual initiative and individual force. The country which stamps out individuality will soon live in the mass alone.

A French writer has claimed that the decay of religious spirit in France is connected with the growth of religious orders of which celibacy is a prominent feature. If religious men and women leave no descendants, their own spirit, at least, will fail of inheritance. A people careless of religion inherit this trait from equally careless ancestors.

Indiscriminate charity has been a fruitful cause of the survival of the unfit. To kill the strong and to feed the weak is to provide for a progeny of weakness. It is a French writer, again, who says that "Charity creates the misery she tries to relieve; she can never relieve half the misery she creates."

There is to-day in Aosta, in Northern Italy, an asylum for the care and culture of idiots. The crétin and the goitre are assembled there, and the marriage of those who cannot take care of themselves ensures the preservation of their strains of unfitness. By caring devotedly for those who in the stress of life could not live alone for a week, and by caring for their children, generation after generation, the good people of Aosta have produced a new breed of men, who cannot even feed themselves. These are incompetent through selection of degradation, while the "Man of the Hoe" is primitively ineffective.

The growth of the goitre in the valleys of Savoy, Piedmont, and Valais, is

In Peace

itself in large part a matter of selection. The boy with the goitre is exempt from military service. He remains at home to become the father of the family. It is said that at one time the government of Savoy furnished the children of that region with lozenges of iodine, which were supposed to check the abnormal swelling of the thyroid gland, known as the goitre. This disease is a frequent cause of idiocy, or cretinism, as well as its almost constant accompaniment. It is said that the mothers gave the lozenges only to the girls, preferring that the boys should grow up to the goitre rather than to the army. The causes of goitre are obscure, perhaps depending on poor nutrition or on mineral substances in the water. The disease itself is not hereditary, so far as known; but susceptibility to it certainly is. By taking away for outside service those who are resistant, the heredity of tendency to goitrous swelling is fastened on those who remain.

Like these mothers in Savoy was a mother in Germany. Not long since a friend of the writer, passing through a Franconian forest, found a young man lying senseless by the way. It was a young recruit for the army who had got into some trouble with his comrades. They had beaten him and left him lying with a broken head. Carried to his home, his mother fell on her knees and thanked God, for this injury had saved him from the army.

The effect of alcoholic drink on race-progress should be considered in this connection. Authorities do not agree as to the final result of alcohol in race-selection. Doubtless, in the long run, the drunkard will be eliminated; and perhaps certain authors are right in regarding this as a gain to the race. On the other hand, there is great force

In Peace

in Dr. Amos G. Warner's remark, that of all caustics gangrene is the most expensive. The people of southern Europe are relatively temperate. They have used wine for centuries, and it is thought by Archdall Reid and others that the cause of their temperance is to be found in this long use of alcoholic beverages. All those with vitiated or uncontrollable appetites have been destroyed in the long experience with wine, leaving only those with normal tastes and normal ability of resistance. The free use of wine is, therefore, in this view, a cause of final temperance, while intemperance rages only among those races which have not long known alcohol, and have not become by selection resistant to it. The savage races which have never known alcohol are even less resistant, and are soonest destroyed by it.

In all this there must be a certain

element of truth. The view, however, ignores the evil effect on the nervous system of long-continued poisoning, even if the poison be only in moderate amounts. The temperate Italian, with his daily semi-saturation, is no more a normal man than the Scotch farmer with his occasional sprees. The nerve disturbance which wine effects is an evil, whether carried to excess in regularity or irregularity. We know too little of its final result on the race to give certainty to our speculations. It is, moreover, true that most excess in the use of alcohol is not due to primitive appetite. It is drink which causes appetite, and not appetite which seeks for drink. In a given number of drunkards but a very few become such through inborn appetite. It is influence of bad example, lack of courage, false idea of manliness, or some defect in character or misfortune in environment which leads to the first steps in drunkenness. The taste once established takes care of itself. In earlier times, when the nature of alcohol was unknown and total abstinence was undreamed of, it was the strong, the boisterous, the energetic, the apostle of "the strenuous life," who carried all these things to excess. The wassail bowl, the bumper of ale, the flagon of wine,— all these were the attribute of the strong. We cannot say that those who sank in alcoholism thereby illustrated the survival of the fittest. Who can say that, as the Latin races became temperate, they did not also become docile and weak? In other words, considering the influence of alcohol alone, unchecked by an educated conscience, we must admit that it is the strong and vigorous, not the weak and perverted, that are destroyed by it. At the best, we can only say that alcoholic selection is a complex force, which makes for temperance — if at all, at a fearful cost of life which without alcoholic temptation would be well worth saving. We cannot easily, with Mr. Reid, regard alcohol as an instrument of race-purification, nor believe that the growth of abstinence and prohibition only prepares the race for a future deeper plunge into dissipation. If France, through wine, has grown temperate, she has grown tame. "New Mirabeaus," Carlyle tells us, " one hears not of; the wild kindred has gone out with this, its greatest." This fact, whatever the cause, is typical of great, strong, turbulent men who led the wild life of Mirabeau because they knew nothing better.

The concentration of the energies of France in the one great city of Paris is again a potent agency in the impoverishment of the blood of the rural

In Peace

districts. All great cities are destroyers of life. Scarcely one would hold its own in population or power, were it not for the young men of the farms. In such destruction, Paris has ever taken the lead. The education of the middle classes in France is almost exclusively a preparation for public life. To be an official in a great city is an almost universal ideal. This ideal but few attain, and the lives of the rest are largely wasted. Not only the would-be official, but artist, poet, musician, physician, or journalist, seeks his career in Paris. A few may find it. The others, discouraged by hopeless effort or vitiated by corrosion, faint and fall. Every night some few of these cast themselves into the Seine. Every morning they are brought to the morgue behind the old Church of Notre Dame. It is a long procession and a sad one from the provincial village to the strife and pitfalls of the great city, from hope and joy to absinthe and the morgue. With all its pitiful aspects the one which concerns us is the steady drain on the life-blood of the nation, its steady lowering of the average of the parent stock of the future.

But far more potent for evil to the race than all these influences, large and small, is the one great destroyer,— War. War for glory, war for gain, war for dominion, its effect is the same, whatever its alleged purpose.

II
IN WAR

II

In War.

Not long ago I visited the town of Novara, in northern Italy. There, in a wheat-field, the farmers have ploughed up skulls of men till they have piled up a pyramid ten or twelve feet high. Over this pyramid some one has built a canopy to keep off the rain. These were the skulls of young men of Savoy, Sardinia, and Austria,— men of eighteen to thirty-five years of age, without physical blemish so far as may be,— peasants from the farms and workmen from the shops, who met at Novara to kill each other over a matter in which they had very little concern. Should the Prince of Savoy sit on his unstable throne or yield it to some one else, this was the question. It matters not the decision. History doubtless re-

The Blood of the Nation

cords it, as she does many matters of less moment. But this fact concerns us,— here in thousands they died. Farther on, Frenchmen, Austrians, and Italians fell together at Magenta, in the same cause. You know the color that we call Magenta, the hue of the blood that flowed out under the olive-trees. Go over Italy as you will, there is scarcely a spot not crimsoned by the blood of France, scarcely a railway station without its pile of French skulls. You can trace them across to Egypt, to the foot of the Pyramids. You will find them in Germany,— at Jena and Leipzig, at Lützen and Bautzen and Austerlitz. You will find them in Russia, at Moscow; in Belgium, at Waterloo. "A boy can stop a bullet as well as a man," said Napoleon; and with the rest are the skulls and bones of boys, "ere evening to be trodden like the grass." "Born to be

In War

food for powder" was the grim epigram of the day, summing up the life of the French peasant. Read the dreary record of the glory of France, the slaughter at Waterloo, the wretched failure of Moscow, the miserable deeds of Sedan, the waste of Algiers, the poison of Madagascar, the crimes of Indo-China, the hideous results of barrack vice and its entail of disease and sterility, and you will understand the "Man of the Hoe." The man who is left, the man whom glory cannot use, becomes the father of the future men of France. As the long-horn cattle reappear in a neglected or abused herd of Durhams, so comes forth the aboriginal man, the "Man of the Hoe," in a wasted race of men.

A recent French cartoon pictures the peasant of a hundred years ago ploughing in a field, a gilded marquis on his back, tapping his gilded snuff-box.

The Blood of the Nation

Another cartoon shows the French peasant of to-day, still at the plough. On his back is an armed soldier who should be at another plough, while on the back of the soldier rides the second burden of Shylock the money-lender, more cruel and more heavy even than the dainty marquis of the old régime. So long as war remains, the burden of France cannot be shifted.

In the loss of war we count not alone the man who falls or whose life is tainted with disease. There is more than one in the man's life. The bullet that pierces his heart goes to the heart of at least one other. For each soldier has a sweetheart; and the best of these die, too,— so far as the race is concerned,— if they remain single for his sake.

In the old Scottish ballad of the "Flower of the Forest" this thought is set forth:—

In War

"I've heard the lilting at each ewe-milking,
  Lassies a-lilting before the dawn of day.
  But now they are moaning on ilka green loaning,
  For the "Flower of the Forest" is a' wed away."

Ruskin once said that "war is the foundation of all high virtues and faculties of men." As well might the maker of phrases say that fire is the builder of the forest, for only in the flame of destruction do we realize the warmth and strength that lie in the heart of oak. Another writer, Hardwick, declares that "war is essential to the life of a nation; war strengthens a nation morally, mentally and physically." Such statements as these set all history at defiance. War can only waste and corrupt. "All war is bad," says Benjamin Franklin, "some only worse than others." "War has its origin in the evil passions of men," and even when unavoidable or

The Blood of the Nation

righteous, its effects are most forlorn. The final effect of each strife for empire has been the degradation or extinction of the nation which led in the struggle.

Greece died because the men who made her glory had all passed away and left none of their kin and therefore none of their kind. "'Tis Greece, but living Greece no more"; for the Greek of to-day, for the most part, never came from the loins of Leonidas or Miltiades. He is the son of the stable-boys and scullions and slaves of the day of her glory, those of whom imperial Greece could make no use in her conquest of Asia. "Most of the old Greek race," says Mr. W. H. Ireland, "has been swept away, and the country is now inhabited by persons of Slavonic descent. Indeed, there is strong ground for the statement that there was more of the old heroic blood of

In War

Hellas in the Turkish army of Edhem Pasha than in the soldiers of King George, who fled before them three years ago." King George himself is only an alien placed on the Grecian throne to suit the convenience of the outside powers, which to the ancient Greeks were merely factions of barbarians. In the late war some poet, addressing the spirit of ancient Greece, appealed to her,—

> "Of all thy thousands grant us three
> To make a new Thermopylæ."

But there were not even three — not even one —"to make another Marathon," and the Turkish troops swept over the historic country with no other hindrance than the effortless deprecation of Christendom.

Why did Rome fall? It was not because untrained hordes were stronger than disciplined legions. It was not that she grew proud, luxurious, corrupt, and thereby gained a legacy of physical weakness. We read of her wealth, her extravagance, her indolence and vice; but all this caused only the downfall of the enervated, the vicious, and the indolent. The Roman legions did not riot in wealth. The Roman generals were not all entangled in the wiles of Cleopatra.

"The Roman Empire," says Seeley, "perished for want of men." You will find this fact on the pages of every history, though few have pointed out war as the final and necessary cause of the Roman downfall. In his recent noble history of the "Downfall of the Ancient World" ("Der Untergang der Antiken Welt," 1897), Professor Otto Seeck, of Greifeswald, makes this fact very apparent. The cause of the fall of Rome is found in the "extinction of the best" ("*Die Ausrottung der*

In War

Besten"), and all that remains to the historian is to give the details of this extermination. He says, "In Greece a wealth of spiritual power went down in the suicidal wars." In Rome "Marius and Cinna slew the aristocrats by hundreds and thousands. Sulla destroyed no less thoroughly the democrats, and whatever of noble blood survived fell as an offering to the proscription of the triumvirate." "The Romans had less of spontaneous power to lose than the Greeks, and so desolation came to them all the sooner. He who was bold enough to rise politically was almost without exception thrown to the ground. *Only cowards remained, and from their brood came forward the new generations.* Cowardice showed itself in lack of originality and slavish following of masters and traditions." Had the Romans been still alive, the Romans of the old republic, neither inside nor outside forces could have worked the fall of Rome. But the true Romans passed away early. Even Cæsar notes the "dire scarcity of men" ("δεινήν ὀλιγανθρωπίαν"). Still there were always men in plenty, such as they were. Of this there is abundant testimony. Slaves and camp-followers were always in evidence. It was the men of strength and character, "the small farmers," the "hardy dwellers on the flanks of the Apennines," who were gone.

"The period of the Antonines was a period of sterility and barrenness. The human harvest was bad." Augustus offered bounties on marriage until "celibacy became the most comfortable and most expensive condition of life." "Marriage," says Metellus, "is a duty which, however painful, every citizen ought manfully to discharge."

"The mainspring of the Roman army," says Hodgkin, "for centuries

In War

had been the patient strength and courage, capacity for enduring hardships, instinctive submission to military discipline, of the population which lined the ranges of the Apennines."

Berry states that an "effect of the wars was that the ranks of the small farmers were decimated, while the number of slaves who did not serve in the army multiplied." Thus "*Vir gave place to Homo*," real men to mere human beings.

With the failure of men grew the strength of the mob, and of the emperor, its exponent. "The little finger of Constantine was stronger than the loins of Augustus." At the end "the barbarians settled and peopled the Roman Empire rather than conquered it." "The Roman world would not have yielded to the barbaric, were it not decidedly inferior in force." Through the weakness of men the emperor assumed divine right. Dr. Zumpt says: "Government, having assumed godhead, took at the same time the appurtenances of it. Officials multiplied. Subjects lost their rights. Abject fear paralyzed the people, and those that ruled were intoxicated with insolence and cruelty."

"The Emperor," says Professor Seeley, "possessed in the army an overwhelming force, over which citizens had no influence, which was totally deaf to reason or eloquence, which had no patriotism because it had no country, which had no humanity because it had no domestic ties." "There runs through Roman literature a brigand's and a barbarian's contempt for honest industry." "The worst government is that which is most worshipped as divine."

So runs the word of the historian. The elements are not hard to find,— extinction of manly blood, extinction

In War

of freedom of thought and action, increase of wealth gained by plunder, loss of national existence.

So fell Greece and Rome, Carthage and Egypt, the Arabs and the Moors, because, their warriors dying, the nation bred real men no more. The man of the strong arm and the quick eye gave place to the slave, the pariah, the man with the hoe, whose lot changes not with the change of dynasties.

Other nations of Europe may furnish illustrations in greater or less degree. Germany guards her men, and reduces the waste of war to a minimum. She is "military, but not warlike"; and this distinction means a great deal from the point of view of this discussion. In modern times the greatest loss of Germany has been not from war, but from emigration. If the men who have left Germany are of higher type than those who remain at home, then the blood of

The Blood of the Nation

the nation is impoverished. That this is the case the Germans in Germany are usually not willing to admit. On the other hand, those competent to judge the German-American find no type of men in the Old World his mental or physical superior.

The tendency of emigration, whether to cities or to other countries, is to weaken the rural population. An illustration of the results of checking this form of selection is seen in the Bavarian town of Oberammergau. This little village, with a population not exceeding fifteen hundred, has a surprisingly large number of men possessing talent, mental and physical qualities far above the average even in Germany. The cause of this lies in the Passion Play, for which for nearly three centuries Oberammergau has been noted. The best intellects and the noblest talents that arise in the town find full scope for

In War

their exercise in this play. Those who are idle, vicious, or stupid are excluded from it. Thus, in the long run, the operation of selection is to retain those whom the play can use and to exclude all others. To weigh the force of this selected heredity, we have only to compare the quality of Oberammergau with that of other Bavarian towns, as, for example, her sister village of Unterammergau, some two miles lower down, in the same valley.

Switzerland is the land of freedom, the land of peace. But in earlier times some of the thrifty cantons sent forth their men as hireling soldiers to serve for pay under the flag of whomsoever might pay their cost. There was once a proverb in the French Court, "*Pas d'argent, pas de Suisses*" (No money, no Swiss); for the agents of the free republic drove a close bargain.

In Lucerne stands one of the noblest

monuments in all the world, the memorial of the Swiss guard of Louis XVI., killed by the mob at the palace of Versailles. It is carved in the solid rock of a vertical cliff above a great spring in the outskirts of the city,— a lion of heroic size, a spear thrust through its body, guarding in its dying paws the Bourbon lilies and the shield of France. And the traveller, Carlyle tells us, should visit Lucerne and her monument, "not for Thorwaldsen's sake alone, but for the sake of the German *Biederkeit* and *Tapferkeit*, the valor which is worth and truth, be it Saxon, be it Swiss."

Beneath the lion are the names of those whose devotion it commemorates. And with the thought of their courage comes the thought of the pity of it, the waste of brave life in a world that has none too much. It may be fancy, but it seems to me that, as I go about in Switzerland, I can distinguish by the

In War

character of the men who remain those cantons who sent forth mercenary troops from those who kept their own for their own upbuilding. Perhaps for other reasons than this Lucerne is weaker than Graubünden, and Unterwalden less virile than little Appenzell. In any event, the matter is worthy of consideration; for this is absolutely certain,— just in proportion to its extent and thoroughness is military selection a cause of decline.

Holland has become a nation of old men, rich, comfortable, and unprogressive. Her sons have died in the fields of Java, the swamps of Achin, wherever Holland's thrifty spirit has built up nations of slaves. It is said that Batavia alone has a million of Dutch graves. The armies of Holland to-day are recruited in every port. Dutch blood is too precious to be longer spilled in her enterprises.

The Blood of the Nation

Spain died of empire centuries ago. She has never crossed our path. It was only her ghost which walked at Manila and Santiago. In 1630 the Augustinian friar La Puente thus wrote of the fate of Spain: "Against the credit for redeemed souls I set the cost of armadas and the sacrifice of soldiers and friars sent to the Philippines. And this I count the chief loss; for mines give silver, and forests give timber, but only Spain gives Spaniards, and she may give so many that she may be left desolate, and constrained to bring up strangers' children instead of her own." "This is Castile," said a Spanish knight; "she makes men and wastes them." "This sublime and terrible phrase," says Lieutenant Carlos Gilman Calkins, from whom I have received both these quotations, "sums up Spanish history."

The warlike nation of to-day is the decadent nation of to-morrow. It has

In War

ever been so, and in the nature of things it must ever be.

In his charming studies of "Feudal and Modern Japan," Mr. Arthur Knapp returns again and again to the great marvel of Japan's military prowess after more than two hundred years of peace. It is astonishing to him that, after more than six generations in which physical courage has not been demanded, these virile virtues should be found unimpaired. We can readily see that this is just what we should expect. In times of peace there is no slaughter of the strong, no sacrifice of the courageous. In the peaceful struggle for existence there is a premium placed on these virtues. The virile and the brave survive. The idle, weak, and dissipated go to the wall. If after two hundred years of incessant battle Japan still remained virile and warlike, that would indeed be the marvel. But that marvel no na-

The Blood of the Nation

tion has ever seen. It is doubtless true that warlike traditions are most persistent with nations most frequently engaged in war. But the traditions of war and the physical strength to gain victories are very different things. Other things being equal, the nation which has known least of war is the one most likely to develop the "strong battalions" with whom victory must rest.

What shall we say of England and her hundred petty wars "smouldering" in every part of the globe?

Statistics we have none, and no evidence of tangible decline that Englishmen will not indignantly repudiate. Besides, in the struggle for national influences, England has had many advantages which must hide or neutralize the waste of war. In default of facts unquestioned, we may appeal to the poets, letting their testimony as to the

In War

reversal of selection stand for what it is worth. Kipling tells us of the cost of the rule of the sea : —

"We have fed our sea for a thousand years,
    And she calls us, still unfed ;
  Though there's never a wave of all her waves
    But marks our English dead.

"If blood be the price of admiralty,
    Lord God, we have paid it in full."

Again, referring to dominion on land, he says : —

"Walk wide of the widow of Windsor,
    For half of creation she owns,
  We've bought her the same with the sword
    and the flame,
  And we've salted it down with our bones.
  Poor beggars, it's blue with our bones."

Finer than this are the lines in the "Revelry of the Dying," written by a British officer, Bartholomew Dowling, it is said, who died in the plague in India : —

The Blood of the Nation

"Cut off from the land that bore us,
  Betrayed by the land we find,
When the brightest are gone before us,
  And the dullest are left behind.
So stand to your glasses steady,
  Tho' a moment the color flies;
Here's a cup to the dead already
  And huzza for the next that dies!"

The stately "Ave Imperatrix" of Oscar Wilde, the last flicker of dying genius in his wretched life, contains lines that ought not to be forgotten:—

"O thou whose wounds are never healed,
  Whose weary race is never run,
O Cromwell's England, must thou yield
  For every foot of ground a son?

"What matter if our galleys ride
  Pine-forest-like on every main?
Ruin and wreck are at our side,
  Stern warders of the house of pain.

"Where are the brave, the strong, the fleet,
  The flower of England's chivalry?
Wild grasses are their winding-sheet,
  And sobbing waves their threnody.

In War

"Peace, peace, we wrong our noble dead
 To vex their solemn slumber so ;
*But childless and with thorn-crowned head
 Up the steep road must England go !*"

We have here the same motive, the same lesson, which Byron applies to Rome : —

"The Niobe of Nations — there she stands,
 Crownless and childless in her voiceless woe,
An empty urn within her withered hands,
 Whose sacred dust was scattered long ago !"

It suggests the inevitable end of all empire, of all dominion of man over man by force of arms. More than all who fall in battle or are wasted in the camps, the nation misses the " fair women and brave men " who should have been the descendants of the strong and the manly. If we may personify the spirit of the nation, it grieves most not over its " unreturning brave," but over those who might have been, but never were,

The Blood of the Nation

and who, so long as history lasts, can never be.

Against this view is urged the statement that the soldier is not the best, but the worst, product of the blood of the English nation. Tommy Atkins comes from the streets, the wharves, the graduate of the London slums, and if the empire is "blue with his bones," it is, after all, to the gain of England that her better blood is saved for home consumption, and that, as matters are, the wars of England make no real drain of English blood.

In so far as this is true, of course the present argument fails. If war in England is a means of race improvement, the lesson I would read does not apply to her. If England's best do not fall on the field of battle, then we may not accuse war of their destruction. The fact could be shown by statistics. If the men who have fallen in England's

In War

wars, officers and soldiers, rank and file, are not on the whole fairly representative of "the flower of England's chivalry," then fame has been singularly given to deception. We have been told that the glories of Blenheim, Trafalgar, Waterloo, Majuba Hill, were won by real Englishmen. And this, in fact, is the truth. In every nation of Europe the men chosen for the army are above the average of their fellows. The absolute best doubtless they are not, but still less are they the worst. Doubtless, too, physical excellence is more considered than moral or mental strength; and certainly, again, the more noble the cause, the more worthy the class of men who will risk their lives for it.

Not to confuse the point by modern instances, it is doubtless true that better men fell on both sides when "Kentish Sir Byng stood for the King" than when the British arms forced the opium trade on China. No doubt, in our own country better men fell at Bunker Hill or Cowpens than at Cerro Gordo or Chapultepec. The lofty cause demands the lofty sacrifice.

It is the shame of England that most of her many wars in our day have cost her very little. They have been scrambles of the mob or with the mob, not triumphs of democracy.

There was once a time when the struggles of armies resulted in a survival of the fittest, when the race was indeed to the swift and the battle to the strong. The invention of "villanous gunpowder" has changed all this. Except the kind of warfare called guerilla, the quality of the individual has ceased to be much of a factor. The clown can shoot down the hero, and "doesn't have to look the hero in the face as he does so." The

In War

shell destroys the clown and hero alike, and the machine gun mows down whole ranks impartially. There is little play for selection in modern war save what is shown in the process of enlistment.

America has grown strong with the strength of peace, the spirit of democracy. Her wars have been few. Were it not for the mob spirit, they would have been still fewer; but in most of them she could not choose but fight. Volunteer soldiers have swelled her armies, men who went forth of their own free will, knowing whither they were going, believing their acts to be right, and taking patiently whatever the fates might hold in store.

The feeling for the righteousness of the cause, "with the flavor of religion in it," says Charles Ferguson, "has made the volunteer the mighty soldier he has always been since the days of Naseby and Marston Moor." Only with volunteer soldiers can democracy go into war. When America fights with professional troops, she will be no longer America. We shall then be, with the rest of the militant world, under mob rule. "It is the mission of democracy," says Ferguson again, " to put down the rule of the mob. In monarchies and aristocracies it is the mob that rules. It is puerile to suppose that kingdoms are made by kings. The king could do nothing if the mob did not throw up its cap when the king rides by. The king is consented to by the mob because of that which in him is mob-like. The mob loves glory and prizes. So does the king. If he loved beauty and justice, the mob would shout for him while the fine words were sounding in the air; but he could never celebrate a jubilee or establish a dynasty. When the crowd gets ready to demand justice and beauty,

In War

it becomes a democracy, and has done with kings."

It was at Lexington that "the embattled farmers" "fired the shot heard round the world." To them life was of less value than a principle, the principle written by Cromwell on the statute book of Parliament: "All just powers under God are derived from the consent of the people." Since this war many patriotic societies have arisen, finding their inspiration in personal descent from those who fought for American independence. The assumption, well justified by facts, is that these were a superior type of men, and that to have had such names in our personal ancestry is of itself a cause for thinking more highly of ourselves. In our little private round of peaceful duties we feel that we might have wrought the deeds of Putnam and Allen, of Marion and Greene, of our Revolu-

The Blood of the Nation

tionary ancestors, whoever they may have been. But if those who survived were nobler than the mass, so also were those who fell. If we go over the record of brave men and wise women whose fathers fought at Lexington, we must think also of the men and women who shall never be, whose right to exist was cut short at this same battle. It is a costly thing to kill off men, for in men alone can national greatness consist.

But sometimes there is no other alternative. It happened once that for "every drop of blood drawn by the lash another must be drawn by the sword." It cost us a million of lives to get rid of slavery. And this million, North and South, was the "best that the nation could bring." North and South, the nation was impoverished by the loss. The gaps they left are filled to all appearance. There are

In War

relatively few of us left to-day in whose hearts the scars of forty years ago are still unhealing. But a new generation has grown up of men and women born since the war. They have taken the nation's problems into their hands, but theirs are hands not so strong or so clean as though the men that are stood shoulder to shoulder with the men that might have been. The men that died in "the weary time" had better stuff in them than the father of the average man of to-day.

Read again Brownell's rhymed roll of honor, and we shall see its deeper meaning:—

> "Allen, who died for others,
>   Bryan of gentle fame,
> And the brave New England brothers
>   Who have left us Lowell's name;
> Bayard, who knew not fear,
>   True as the knight of yore,
> And Putnam and Paul Revere,
>   Worthy the names they bore;

The Blood of the Nation

Wainwright, steadfast and true,
    Rodgers of brave sea-blood,
And Craven, with ship and crew,
    Sunk in the salt-sea flood;
Terrill, dead where he fought,
    Wallace, that would not yield;
Sumner, who vainly bought
    A grave on the foughten field,
But died ere the end he saw,
    With years and battles outworn;
There was Harmon of Kennesaw,
And Ulric Dahlgren, and Shaw
    That slept with his Hope Forlorn;
Lytle, soldier and bard,
    And the Ellets, sire and son;
Ransom, all grandly scarred,
And Redfield, no more on guard.
    But Alatoona is won!"

So runs the record, page after page: —

    "All such, and many another,
        Ah! list how long to name!"

And these were the names of the officers only. Not less worthy were the men in the ranks. It is the paradox of democracy that its greatness is chiefly in the

In War

ranks. "Are all the common men so grand, and all the titled ones so mean?"

North or South, it was the same. "Send forth the best ye breed" was the call on both sides alike, and to this call both sides alike responded. As it will take "centuries of peace and prosperity to make good the tall statures mowed down in the Napoleonic wars," so like centuries of wisdom and virtue are needed to restore to our nation its lost inheritance of patriotism,— not the capacity for patriotic talk, for of that there has been no abatement, but of that faith and truth which "on war's red touchstone rang true metal." With all this we can never know how great is our real misfortune, nor see how much the men that are fall short of the men that ought to have been.

It will be said that all this is exaggeration, that war is but one influence among many, and that each and all of

The Blood of the Nation

these forms of destructive selection may find its antidote. This is very true. The antidote is found in the spirit of democracy, and the spirit of democracy is the spirit of peace. Doubtless these pages constitute an exaggeration. They were written for that purpose. I would show the " ugly, old, and wrinkled truth stripped clean of all the vesture that beguiles." To see anything clearly and separately is to exaggerate it. The naked truth is always a caricature unless clothed in conventions, fragments taken from lesser truths. The moral law is an exaggeration: " The soul that sinneth, it shall die." Doubtless one war will not ruin a nation. Doubtless it will not destroy its virility or impair its blood. Doubtless a dozen wars may do all this. The difference is one of degree alone; I wish only to point out the tendency. That the death of the strong is a true cause of the decline of nations is a fact

In War

beyond cavil or question. The "man who is left" holds always the future in his grasp. One of the great books of our new century will be some day written on the selection of men, the screening of human life through the actions of man and the operation of the institutions men have built up. It will be a survey of the stream of social history, its whirls and eddies, rapids and still waters, and the effect of each and all of its conditions on the heredity of men. The survival of the fit and the unfit in all degrees and conditions will be its subject-matter. This book will be written, not roughly and hastily, like the present fragmentary essay, still less will it be a brilliant effort of some analytical imagination. It will set down soberly and statistically the array of facts which as yet no one possesses; and the new Darwin whose work it shall be must, like his predecessor, spend twenty-five years in the gathering of "all facts that can possibly bear on the question." When such a book is written, we shall know for the first time the real significance of war.

If any war is good, civil war must be best. The virtues of victory and the lessons of defeat would be kept within the nation. This would protect the nation from the temptation to fight for gold or trade. Civil war under proper limitations could remedy this. A time limit could be adopted, as in football, and every device known to the arena could be used to get the good of war and to escape its evils.

For example, of all our States, New York and Illinois have doubtless suffered most from the evils of peace, if peace has evils which disappear with war. They could be pitted against each other, while the other States looked on. The "dark and bloody ground" of

In War

Kentucky could be made the arena. This would not interfere with trade in Chicago, nor soil the streets in Baltimore. The armies could be filled up from the ranks of the unemployed, while the pasteboard heroes of the national guard could act as officers. All could be done in decency and order, with no recriminations and no oppression of an alien foe. We should have all that is good in war, its pomp and circumstance, the "grim resolution of the London clubs," without war's long train of murderous evils. Who could deny this? And yet who could defend it?

If war is good, we should have it regardless of its cost, regardless of its horrors, its sorrows, its anguish, havoc, and waste.

But it is bad, only to be justified as the last resort of "mangled, murdered liberty," a terrible agency to be evoked

The Blood of the Nation

only when all other arts of self-defence shall fail. The remedy for most ills of men is not to be sought in " whirlwinds of rebellion that shake the world," but in peace and justice, equality among men, and the cultivation of those virtues we call Christian, because they have been virtues ever since man and society began, and will be virtues still when the era of strife is past and the " redcoat bully in his boots " no longer " hides the march of man from us."

It is the voice of political wisdom which falls from the bells of Christmastide: "Peace on earth, good will towards men!"

Printed in the USA
CPSIA information can be obtained
at www.ICGtesting.com
LVHW041459280724
786735LV00008B/123